Building a Safe Room

Step-by-Step Instructions for Design, Construction, and Equipping Your Residential Safe Haven

Richard Man

Contents

INTRODUCTION

"Safety is a right, not a privilege—a fortified room in your home can be your ultimate peace of mind."

In the quest for security, the concept of a safe room has transformed from a luxury feature in the dwellings of the elite to a practical necessity in the modern home. This transformation tells a compelling story of adaptation and survival, where architecture meets personal safety.

Understanding the Need for a Safe Room

The history of safe rooms is as old as the history of conflict and natural disasters themselves. Originally known as panic rooms, they were fortified spaces where nobility could retreat to in times of siege. Today, they serve a similar purpose, offering sanctuary in the face of unforeseen threats, be they man-made or natural.

Recent statistics paint a stark picture of increasing natural disasters and home invasions, highlighting the unpredictable nature of our world. For instance, the uptick in severe weather events has led many to consider the security of their homes against forces that spare no one.

Similarly, urbanization has seen a rise in home invasions, turning the safe room into a vital refuge in worst-case scenarios.

Personal stories of safe room effectiveness resonate with a profound sense of relief and gratitude. There are countless accounts of families who have weathered hurricanes or deflected home invasions, all thanks to the sanctuary of a well-prepared safe room. These testimonies not only underscore the value of such a space but also provide real-world insights into their practical benefits.

What This Book Will Cover

This book is a pragmatic guide aimed at demystifying the process of designing, constructing, and equipping a safe room within your residence. It's a detailed blueprint for those who recognize the importance of having a dedicated space that ensures the safety and security of their loved ones.

We will begin by addressing common questions and concerns, providing you with the knowledge to make informed decisions. Whether you're a homeowner looking to retrofit a space or you're planning to integrate a safe room into a new construction, this guide will serve as an indispensable resource.

The significance of meticulous planning cannot be overstressed. A safe room is only as effective as the thought and effort put into its creation. From the placement of a wall to the choice of a communication device, every detail holds the potential to save lives.

In the following chapters, we will explore each aspect of safe room creation with an eye for actionability. You will not only learn the facts but also how to apply them. By the end of this book, you will be equipped with the expertise to build a safe haven — a testament to the resilience and foresight that defines preparedness in the 21st century.

THE BLUEPRINT OF SAFETY

"A well-planned safe room is a fortress of solitude in times of chaos."

Alright, let's talk about picking out the perfect spot for your safe room. You know, it's kind of like choosing where to plant a tree. You wouldn't put it somewhere the roots could get damaged or where it wouldn't get enough sunlight, right? Similarly, with a safe room, you need a spot that's going to stand strong and remain unshakable, no matter what's howling outside.

First off, you've got to think like a detective looking for clues in your own home. Which walls are the sturdiest? Is there a place that's naturally shielded from the elements or potential threats? Basements often come up as a top choice because they're below ground level, which naturally gives them a leg up against things like storms and high winds. But, let's not forget, you've got to consider flooding risks down there, too.

Now, if you're not keen on heading underground, maybe you're looking at a closet or a room without windows in the core of your house. These spots are typically more protected and, because they're surrounded by other parts of your house, they've got a bit of a buffer zone, which is great for insulation and soundproofing as well.

The thing is, no matter where you look, you've got to consider the bones of the place. Are the walls solid? Are they made of concrete or reinforced in some way? And the floors – are they going to hold up if something heavy comes crashing down?

It's not just about the now, either. You need to be thinking long-term. Is this place going to stay sturdy over the years? You don't want to be redoing this in a decade because the foundation wasn't as solid as you thought.

And hey, while you're sizing up your space, don't forget to think about access. You want to be able to get to your safe room quickly if you need to. So, a spot that's too out of the way might not be your best bet, even if it's built like a fortress.

So, grab your magnifying glass (figuratively, of course) and take a good, hard look at your space. And if you're not quite sure about it, it's a smart move to bring in a structural engineer who can give you the lowdown on the best spot for your safe haven. Remember, when it comes to a safe room, solid ground is your best friend.

***

Accessibility during an emergency

Alright, let's chat about getting to your safe room when the going gets tough. Imagine it's like the middle of the night, and you're all cozy in bed. Suddenly, you hear that dreaded siren or your phone blasts an emergency alert. Your heart's racing, and you need to move – and fast.

This is where the "where" of your safe room really comes into play. You want a spot that's quick to get to, no matter where you are in the house. Think about those middle-of-the-night drills. You shouldn't need to navigate a maze or dash up three flights of stairs to reach safety. Ideally, it's a straight shot from wherever you spend the most time.

Also, think about everyone in your home. Got kids? Elderly folks? Someone with a disability? The path to your safe room needs to work for them, too. Wide doorways, ramps if needed – the works. It's like when you're planning a party, and you make sure everyone can get in and find the fun without a hitch.

And, you know, emergencies love to mess with you. Power could be out, meaning no lights, and things could be falling off shelves. So, you'll want this route to your safe room to be as clear and uncluttered as possible. No tripping over that funky rug or bumping into the side table.

Now, let's not forget about a back-up plan. Say the usual way is blocked – you need a plan B. Maybe a secondary route that's also easy to navigate, just in case. It's like when you're driving to your favorite restaurant and the main road is closed, you've got that sneaky back road that gets you there just as easy.

Bottom line: you want everyone to get to that safe room quickly and safely, no matter what's going down. It's all about peace of mind, knowing you can get there when every second counts.

***

Considering the Proximity to Utilities and Escape Routes

let's dive into the nitty-gritty of picking a spot for your safe room in relation to the life-support stuff - utilities - and your quick getaway options - escape routes.

You know how when you're picking a campsite, you want to be close to the water and have a clear path out just in case a bear decides to join you? That's kind of how you need to think about your safe room. You want it close enough to utilities so you can hook up to power, water, and maybe even the internet without running cables all over the place. Because let's face it, in an emergency, the last thing you want is to trip over a bunch of wires just to charge your phone or get some lights on.

But it's a balancing act. You don't want to be too close to things like gas lines or anything that could become a hazard if they were to, say, burst or break during a disaster. It's a bit like not pitching your tent under a giant, old tree. Looks sturdy, but one storm and crunch!

Now, let's chat escape routes. In a perfect world, your safe room would have a secret tunnel that pops you out into freedom like a spy movie, but back in the real world, you want at least a couple of solid exits. These should be easy to open from the inside but secure from the outside, kind of like those emergency exits in theaters - there in case you need them, but not drawing attention to themselves.

Think about where these exits lead, too. It's no good if they dump you right into a worse situation or into the arms of whatever you're taking shelter from. They should offer a clear path out of the house to a safe distance away, or even better, to another secure location.

So, when you're planning this all out, picture it like you're setting up the ultimate game of hide and seek. You want to be hidden away, but when it's time to bolt, you need a clear, safe path to 'home base,' which in real life could be your neighbor's place or down the block out of harm's way.

***

Designing for Durability and Use

Alright, let's dive into the meaty stuff – making sure your safe room is tough as nails and lasts longer than that fruitcake from last Christmas.

Materials that offer the best protection

- **Concrete is King**: When it comes to protection, you want something that can take a hit and not flinch. Reinforced concrete? That's your heavyweight champ. It's like the body armor for your safe room.

- **Steel – The Ironclad Choice**: If you're going for that impenetrable fortress vibe, steel plating is the way to go. Think of it as the superhero suit for your walls – tough against most anything that gets thrown its way.

- **Bullet-Resistant Fiberglass – Silent Guardian**: Now, for those looking to stop a speeding bullet, literally, bullet-resistant fiberglass panels are your silent guardians. They slide into walls like a secret shield.

Design principles for longevity and maintenance

- **Keep it Simple and Strong**: Complexity is the enemy of endurance. The simpler the design, the fewer things that can go wrong. It's like building with Lego blocks – the less fancy the construction, the sturdier it holds.

- **Accessibility for Repairs**: Design so you can get to the important stuff for maintenance. There's no point in having a secure door if you can't fix it when it jams.

- **Weather the Weather**: Whether it's heat, cold, or moisture, make sure your room can stand up to the local climate. It's like dressing for the season, but for your room.

Customizing the design to fit your home's architecture

- **Blend it In**: You don't want your safe room sticking out like a sore thumb. Work with an architect to make sure it blends with the rest of your house, like a secret passageway that feels like part of the original design.

- **Interior Harmony**: The inside should feel like a continuation of your home, not a bunker. Comfort matters, because if you have to hunker down, you might as well do it in style.

- **Resale Value**: Think about the future. A well-designed safe room can be a selling point, not a panic button. It should add value, not questions, to your home.

So there you have it. When you're putting together the ultimate safe room, think of it as tailoring a bespoke suit – it needs to fit perfectly, suit the occasion (or potential disaster), and last for years to come.

***

Securing Permissions and Legal Considerations

Understanding building codes and regulations

- **The Rulebook**: Before you start dreaming up your safe room, you've gotta flip through the local rulebook – the building codes. These are like the referee's rules in sports; they keep things safe and fair.

- **Playing by the Rules**: Get familiar with what's expected in terms of structure, electrical, and ventilation. It's like making sure your car passes inspection – you don't want to be caught off guard.

- **Expert Advice**: When in doubt, bring in the pros. A chat with a building inspector or an architect can save you a ton of headaches later. They're like the coaches that help you play a clean game.

Navigating zoning laws and neighborhood restrictions

- **Location, Location, Regulation**: Zoning laws dictate what you can build and where. Your dream safe room spot might be in a no-go zone. It's like having a perfect spot for a treehouse, only to find out it's in a nature reserve.

- **Keeping Up with the Joneses**: Neighborhood restrictions can be a whole other ballgame. Sometimes, it's not the city you need to worry about, but what the Homeowners Association (HOA) says. They're like the neighborhood watch of what your house can look like.

- **Smooth Sailing**: Keep your neighbors in the loop to avoid complaints. It's better to invite them to the barbecue before you build, rather than after they've called the HOA on you.

The process of obtaining necessary permits

- **Permit Quest**: Think of getting permits like a quest in a video game. You need to collect all the right documents, get them stamped, and pay the gold (fees) to proceed.

- **Inspection Gauntlet**: After you file for permits, you'll face the inspection gauntlet. Inspectors will check to make sure you're not cutting corners. It's like the boss level where they make sure you're up to snuff.

- **The Waiting Game**: Patience is key. Getting permits can be like waiting for your favorite band to start playing – it feels like forever, but it's worth it for the main event.

In short, think of securing permissions and navigating legalities as the pre-game warm-up. It's not the most thrilling part of building your safe room, but without it, you won't even get to play the game. Plus, doing it right means you sleep easy, knowing your safe haven is up to code and fully legal.

THE WALLS THAT PROTECT YOU

"Every wall in your safe room tells a story of resilience."

Choosing Your Materials

Comparing traditional and modern building materials

- **Old School vs. New School**: On one side, you've got classic choices like concrete and brick – they're the old-school bouncers that have been keeping bad stuff out for ages. On the flip side, there are new kids on the block like Kevlar and advanced composites, which are like the high-tech security guards with all the gadgets.

- **Strength Test**: Think about what you're up against. Is it weather, fire, or maybe noise? Each material has its own superpower, like concrete's brute strength or Kevlar's bullet-stopping chops.

- **The Mix-and-Match Approach**: Sometimes, you've got to play matchmaker and combine materials to get the best of both worlds. It's like making a smoothie – a bit of this, a bit of that, and you've got something that's both tasty and nutritious.

Cost vs. protection: Finding the balance

- **Budget Battles**: Your wallet might not be as fortified as your safe room needs to be. It's all about balance – like picking a car that's both safe and doesn't guzzle gas like there's no tomorrow.

- **The Sweet Spot**: You're looking for that Goldilocks zone where the cost of materials meets the level of protection you need. It's like picking the right gear for a hike – enough to be prepared, but not so much that you can't move.

- **Future-Proofing**: Investing a bit more now could save you big time in the future. Think of it as buying the better boots so you don't end up with blisters halfway through the journey.

Sourcing and evaluating suppliers

- **Supplier Sleuthing**: Time to put on your detective hat and dig into where your materials are coming from. Are the suppliers reputable? Are they the 'here today, gone tomorrow' kind, or the 'solid as their concrete' kind?

- **Quality Check**: Just like you wouldn't buy a parachute without a double-check, you need to ensure the materials are up to standard. Look for certifications, quality controls, and customer reviews.

- **Delivery Drama**: Make sure whoever you're buying from can deliver the goods, literally. You don't want your project on pause because your steel plates are stuck in shipping limbo.

Choosing materials for your safe room is like assembling a team for a heist movie – everyone needs to bring something unique to the table, be reliable, and work well together. And remember, the best team might cost a bit more, but they're worth it when it's game time.

***

Construction Techniques for Maximum Security

Detailed steps for constructing resilient walls

- **Laying the Groundwork**: Like baking a cake, you start with a solid recipe – or in this case, blueprints. Make sure they're detailed and tailored to your specific needs.

- **Foundation First**: Your walls are only as good as what they stand on. Ensure your foundation is level, stable, and built to hold the weight – think heavyweight titleholder, not featherweight.

- **Building Up**: Use rebar, mesh, or other reinforcements within your walls – it's like the skeleton that keeps everything upright. Then, pour concrete or stack your chosen materials with care, like a kid with Lego, precision matters.

DIY tips versus professional contracting

- **Know Your Limits**: Sure, you can paint a room or fix a leaky tap, but if you're not Bob the Builder, think twice before DIY-ing a safe room. This isn't a weekend project; it's a fortress of safety.

- **The Pro Path**: Hiring pros isn't admitting defeat; it's playing it smart. They're the A-Team who can navigate the tricky stuff, and they come with warranties – your personal assurance plan.

- **Teamwork for the Win**: If you've got skills and want to be hands-on, work alongside the contractors. It's about teamwork – they'll appreciate the help, and you'll learn a thing or two.

Common pitfalls in wall construction and how to avoid them

- **Shortcut Syndrome**: Taking shortcuts is like skipping leg day – eventually, it'll come back to haunt you. Don't skimp on materials or rush the curing process.

- **Weather Watch**: Mother Nature doesn't pause for construction. Plan around her moods, or you might find your hard work washed away by rain or cracked by the cold.

- **Inspection Oversight**: Skipping inspections or not following codes isn't just illegal; it's dangerous. It's like playing soccer without a referee – chaos will ensue, and someone could get hurt.

Constructing the walls of your safe room isn't something to be taken lightly. It's like fortifying a castle. Every block, every pour, every weld is a line of defense against whatever the world throws your way. So, gear up, plan well, and build strong.

***

Reinforcement Strategies

Upgrading existing walls for enhanced safety

- **Inspection Time**: First things first, get those existing walls checked out. You want to know what you're working with – is it a strong silent type or a bit of a fixer-upper?

- **Layer Up**: Think of your walls like you're dressing for a cold winter day. You want layers – maybe some plywood, then steel sheeting, or even kevlar panels. Each layer adds a bit more warmth – or in this case, protection.

- **Seal the Deal**: Don't forget the little nooks and crannies. Use sealants and reinforcements around windows and doors. It's like weatherproofing your house, but you're crime-proofing it instead.

Incorporating bulletproof and fire-resistant materials

- **Bulletproofing 101**: If you're worried about gunfire, bullet-resistant materials are your best friends. They're like the bodyguards of the wall world – tough and ready to take a hit.

- **Resisting the Heat**: Fire-resistant materials, on the other hand, are like having a firefighter built into your wall. They resist the heat and keep the flames at bay.

- **Combo Platter**: Sometimes you've got to mix and match. A bit of bulletproof here, some fire-resistant there – it's like creating a custom armor for your home.

Using technology for structural monitoring

- **Smart Walls**: These days, walls can be smart too. Install sensors to monitor the integrity of your safe room. It's like having a doctor check your pulse – but for your walls.

- **Alarm Systems**: Set up alarms that trigger if the structural integrity is compromised. It's better to be woken up by a false alarm than to be caught off guard.

- **Regular Check-Ups**: Use tech to schedule regular inspections of your walls. Think of it like a car service – regular check-ups keep it running smoothly.

Reinforcing your safe room isn't just slapping some extra stuff on the walls and calling it a day. It's a thoughtful process of layering, integrating technology, and choosing the right materials for the job. It's giving your safe room the armor it needs to protect you and your loved ones, come what may.

***

Doors, Locks, and Ensuring Access

"A door is only as strong as its lock—and the resolve of those who trust it to protect them."

Selecting a Secure Door

The best materials for high-security doors

- **Steel Stands Strong**: When it comes to doors, steel is like the superhero of materials. It's tough, resistant, and can take a lot of beating. It's your first line of defense.

- **Core Values**: Check out the core of the door. Solid metal cores offer maximum security. It's like the filling in a sandwich - the heartier it is, the better.

- **Composite Materials**: Don't overlook composites. They can offer a blend of steel's strength with other benefits like insulation or fire resistance. It's a bit like a high-tech athletic shoe, designed for performance and durability.

Features to look for in safe room doors

- **Locks that Rock**: A secure door is only as good as its lock. Look for high-grade, tamper-resistant locks. Think multiple deadbolts, biometric systems, or electronic locks that are tough nuts to crack.

- **Hidden Hinges**: Exposed hinges are like an Achilles' heel. Go for internal or reinforced hinges that can't be tampered with from the outside.

- **Peek-a-Boo, I See You**: Consider a peephole or a camera system. It's like having a periscope in a submarine - you need to see what's happening outside without exposing yourself.

Custom versus off-the-shelf options

- **Tailor-Made**: Custom doors can be designed to fit your specific needs and aesthetics. It's like getting a suit tailor-made; it just fits better.

- **Off-the-Shelf Efficiency**: Pre-made doors can be more cost-effective and quicker to install. It's like buying a suit off the rack - it might not fit perfectly, but it does the job.

- **Balance Act**: Weigh the pros and cons of each. Custom doors offer more flexibility in design and features, while off-the-shelf doors are generally more budget-friendly and readily available.

Selecting a secure door for your safe room is like choosing a knight to guard the entrance to your castle. You want it strong, reliable, and equipped with the best armor and weapons to keep the intruders at bay. The right door isn't just a barrier; it's a peace of mind.

***

Locks and Access Control

Mechanical versus electronic lock systems

- **Old School Mechanical**: Mechanical locks are the tried-and-true warriors of the lock world. They don't rely on electricity, so they're like the trusty steed that doesn't need fancy tech to get the job done.

- **Futuristic Electronic**: Electronic locks, on the other hand, are like having a smart assistant for your door. They offer features like keypads or card access, but remember, they do need power to work.

- **Pros and Cons**: It's a bit like choosing between a classic car and a modern electric vehicle. The classic is reliable and straightforward, while the electric offers more bells and whistles but needs charging.

Biometric and multi-factor authentication

- **The Personal Touch**: Biometric systems are like your personal bodyguards. They use fingerprints, retina scans, or facial recognition – something unique to you – to grant access.

- **Layering Security**: Multi-factor authentication is like having a series of checkpoints. You might need a code and a fingerprint, ensuring that even if one layer is breached, you've got another line of defense.

- **Tech-Savvy Security**: This approach is for those who love technology and want that extra level of security. It's like having a high-tech security system – a bit more complex, but it offers peace of mind.

Remote access control and surveillance options

- **Remote Control**: With remote access, you can control your locks from afar, like a wizard with a magic wand. It's handy if you need to let someone in while you're away.

- **Eyes Everywhere**: Surveillance systems can be integrated with your access control, so you can see who's at the door before you even think of unlocking it. It's like having your own command center.

- **Smart Home Integration**: If you're into smart homes, integrating your locks and surveillance with your home system is like linking all your magic spells together. Control everything from your phone or a central panel.

When it comes to locks and access control for your safe room, it's about finding the right mix of old-school reliability and new-age technology. Whether you go for a mechanical system, electronic sophistication, or a blend of both, it's all about making sure that when your door locks, you feel safe and sound inside.

***

Installation and Testing

Step-by-step door installation guide

1. **Measure Twice, Install Once**: Start by accurately measuring your doorway. It's like the tailor measuring you for a suit – precision is key.

2. **Frame It Right**: Ensure the door frame is as strong as the

door itself. It's the foundation – like setting up a good base for a heavy piece of furniture.

3. **Attach and Align**: Fix the door into the frame. This part is crucial – it's like hanging a painting, but imagine the painting weighs a ton and your life depends on it being straight.

4. **Locks and Hardware**: Install the locking mechanisms carefully. Think of it like assembling a high-tech gadget – every piece needs to be in the right place.

5. **Seal and Secure**: Make sure the door is sealed properly to prevent any breaches. It's like waterproofing a boat – no leaks allowed.

Professional installation vs. DIY

- **DIY Pros and Cons**: Going the DIY route can be rewarding and cost-effective if you have the skills. But it's like baking a soufflé – not as easy as it looks and lots can go wrong.

- **Call in the Pros**: Professional installers are like the special forces of door installation. They've done this a thousand times and know all the tricks of the trade.

- **Quality Assurance**: With professionals, you often get guarantees and warranties. It's an extra layer of security, knowing the job's been done right.

Regular maintenance and testing schedules

- **Routine Checks**: Treat your door like a car. It needs regular check-ups to ensure everything's working as it should. Set a schedule for maintenance – maybe every six months.

- **Test the Locks**: Regularly test the locks, hinges, and any electronic systems. It's like a fire drill – you need to know they'll work when you really need them.

- **Keep It Clean**: Simple, but important. Keep the door clean from dust and debris, especially the moving parts. It's like keeping your tools rust-free – they last longer and work better when you take care of them.

Installing and maintaining a secure door isn't just about following instructions. It's about understanding that this door is more than just an entry point; it's a guardian of your safety. So, whether you DIY or bring in the experts, make sure it's done right, and then keep it in top condition. After all, a well-maintained door is a reliable door.

Air to Breathe, Light to See

"Survival isn't just about staying safe, it's about living well under pressure."

Designing a ventilation system for a sealed environment

- **Breathe Easy**: First up, designing ventilation for a sealed room is like planning an underwater air supply. You need enough fresh air without compromising security.

- **Controlled Airflow**: Install vents that allow air in and out but are secure and tamper-proof. Think of them like snorkels – they let you breathe but keep the water out.

- **Circulation is Key**: Ensure the air circulates properly. Stagnant air is a no-go. It's like keeping a fan running in a stuffy room – keeps everything fresh.

Filters and air purifying technology

- **Filter It Out**: Your ventilation should have filters – and we're talking high-grade, like HEPA filters. They catch the tiny bad stuff you can't see, like dust, allergens, or even certain chemicals.

- **Tech to the Rescue**: Consider air purifying technology. There are systems that can zap germs or neutralize chemicals. It's like having a superhero for your air, keeping the villains out.

- **Layer the Protection**: Think about multiple layers of filters. Start with a pre-filter for the big stuff, then get finer as you go. It's like having a security team – each layer catches something the one before might have missed.

Backup systems and fail-safes

- **Plan B for Breathing**: Always have a backup. If your primary system fails, you need another way to get fresh air. It could be as simple as a manual vent.

- **Power Plays**: If your system is powered, have a backup power source. Batteries, generators – it's like having a spare tire, essential for when you hit a bump.

- **Regular Drills**: Test your backup systems regularly. It's like a fire drill – you do it hoping you'll never need it, but you've got to know it works.

Designing a ventilation system for your safe room isn't just about moving air around; it's about creating a lifeline that ensures you can breathe safely, no matter what's happening outside. It's about layering your defenses, from sturdy vents to smart tech, all backed up by fail-safes that keep the air flowing when everything else stops.

***

Lighting Solutions

Emergency lighting options

- **Always Ready**: Emergency lighting is like having a reliable flashlight during a blackout. It automatically kicks in when the power goes out. Think of installing lights that switch to battery power or have a backup generator.

- **Glow in the Dark**: Consider low-level lighting that's always on, like those glow-in-the-dark stickers you had as a kid. They won't illuminate a room, but they'll guide you in pitch darkness.

- **Versatile Fixtures**: Install lights that can be dimmed or brightened depending on the need. It's like having sunglasses and a sun hat – you adjust as the day goes.

Energy-efficient and long-lasting light sources

- **LED All the Way**: LEDs are the champions of efficiency. They use less power and last ages. It's like the energy-saving mode on your phone – does more with less.

- **Smart Controls**: Smart bulbs or systems can save energy by adjusting based on time of day or occupancy. It's like having a clever housekeeper who turns off lights in empty rooms.

- **Quality Counts**: Invest in high-quality bulbs and fixtures. Cheaper ones might save pennies now but can cost dollars later. It's like buying a good pair of shoes – they last longer and perform better.

Installation and maintenance of lighting systems

- **Professional Touch**: Unless you're an electrician, professional installation is the way to go. It's like surgery – best left to the experts.

- **Accessibility is Key**: Install lights where they're easy to maintain. No one wants to be climbing ladders in a crisis.

- **Routine Checks**: Regularly test your lights, especially emergency ones. It's like checking the smoke alarm – a simple step that can be a lifesaver.

Lighting in your safe room is more than just a convenience; it's an integral part of the safety features. It should be reliable, efficient, and easy to manage. Good lighting can not only guide you in an emergency but also create a sense of normalcy in a situation that's anything but normal.

***

Power Supply and Backup

Assessing power needs for your safe room

- **Energy Audit**: Start by figuring out how much juice you'll need. It's like planning for a camping trip – you need to know how much food and water to bring.

- **Essentials First**: List out the must-haves – lighting, ventilation, communication devices. It's like packing your survival kit; prioritize the essentials.

- **Power Play**: Once you know what you need, calculate the total power consumption. This will tell you how big of a power supply or backup you need. Think of it as planning your budget – you need to know what you can afford to run.

Uninterruptible power supplies (UPS) and generators

- **UPS for the Save**: An Uninterruptible Power Supply is like an emergency parachute. It keeps things running smoothly when the main power source drops out.

- **Generator Grit**: Generators are the heavy lifters. They can power your safe room for extended periods. It's like having a backup team ready to jump in when the starters get tired.

- **Fuel for Thought**: Remember, generators need fuel. Store enough, but safely. It's like stocking up on firewood for winter – necessary, but you need to keep it dry and accessible.

Renewable energy options and storage

- **Going Green**: Solar panels or wind turbines can be great renewable options. They're like growing your own veggies – a bit of work to set up, but they provide in the long run.

- **Battery Banks**: Store that green energy in battery banks. They're the pantry of your power supply, keeping that home-grown electricity ready for use.

- **Consistency is Key**: Renewable sources can be less predictable, so having a consistent backup like a generator or grid power is crucial. It's like having a spare tire – essential for when the unexpected happens.

Managing power in your safe room is all about preparation and balance. You need to know what you need, have a reliable way to get it, and always have a backup plan. Think of it as planning a long journey – you need to know your route, have enough fuel to get there, and always have a map for when you need to take a detour.

COMMUNICATION LINES

"When the outside world is silent, your safe room's voice must be loud and clear."

Establishing Robust Communication Channels

Landlines, cell phones, and internet access in a safe room

- **Old School Landlines**: Landlines are like the trusty old bicycle in your garage – not flashy, but they work even when the power's out. In a safe room, having a hard-wired phone can be a lifeline.

- **Cellular Connection**: Cell phones are great, but signal strength can be a gamble. It's like fishing – sometimes you catch a big one, sometimes you get nothing.

- **Wi-Fi Wise**: If you can extend your internet into the safe room, do it. It's like having a digital Swiss Army knife –

you've got access to communication, information, and entertainment.

Satellite phones and ham radios for emergencies

- **Satellite Phones**: These are like having a hotline to the world. They work almost anywhere, making them invaluable when traditional networks are down. It's like sending a carrier pigeon when all the mail trucks have stopped.

- **Ham Radios**: Ham radios are the old-timers of emergency communication, but they're still super effective. They can connect you with a network of helpful operators. It's a bit like having a walkie-talkie with a superhero on the other end.

Ensuring redundancy in communication tools

- **Double Up**: Always have more than one way to communicate. If the cell network goes down, you've got your landline or internet. Think of it like having both a belt and suspenders – it's all about not getting caught with your pants down.

- **Power Backup**: Make sure your communication devices have backup power. Whether it's extra batteries, a hand-crank charger, or a solar charger, you don't want to be left in silence because you ran out of juice.

- **Test Runs**: Regularly test all your communication methods. It's like a fire drill for your phone or radio – you need to know it works before you really need it.

In a nutshell, robust communication in your safe room is about covering all your bases. You want a mix of old and new tech, each with its own power source, so no matter what happens outside, you can still reach out. It's not just about being able to call for help; it's about staying connected to the world.

***

Staying Informed

Best practices for obtaining real-time information

- **Diversify Your Sources**: In today's world, information is like a buffet – you've got lots of options. Don't rely on just one source. Mix it up with news apps, websites, and social media to get a well-rounded view.

- **Alert Systems**: Sign up for local and national emergency alert systems. It's like having an alarm clock for danger – it'll wake you up when something's going down.

- **Stay Current**: Keep your apps and devices updated. Outdated software can be like trying to catch fish with a broken net – you might miss the big ones.

Emergency radio frequencies and information sources

- **Radio Reliability**: A battery-powered or hand-crank emergency radio is a must-have. It's your link to official broadcasts even when other systems are down.

- **Know Your Frequencies**: Familiarize yourself with local emergency frequencies and national weather service channels. It's like knowing the right TV channel for your favorite show.

- **Trusted Networks**: Identify and bookmark trusted news and information networks, both local and national. It's like building a library of the best books – you know they won't let you down.

Preparing for misinformation and communication blackouts

- **Critical Thinking Cap On**: In the age of information overload, not everything you hear or read is true. Be like a detective – question the source and look for confirmation.

- **Blackout Prep**: Have a plan for communication blackouts. This might mean having physical maps, printed emergency contacts, and a list of essential information.

- **Community Ties**: Stay connected with your local community. Sometimes, the best information comes from your neighbors. It's like having a neighborhood watch – everyone looks out for each other.

Staying informed during an emergency is like being a captain in stormy seas. You need your instruments – radios, apps, networks – but you also need to know how to read them and what to do when they go silent. Keeping a level head and a critical eye will help you navigate through the waves of information and misinformation alike.

***

Reaching Out for Help

SOS signals and emergency beacons

- **SOS Basics**: The universal distress signal (three short, three long, three short) can be a lifesaver. It's like having a secret handshake that everyone understands in an emergency.

- **High-Tech Beacons**: Investing in an emergency beacon is like having a panic button that sends your GPS location to rescue services. It's your "break glass in case of emergency" tool.

- **Visual and Audio Signals**: Flares, whistles, or even brightly colored banners can be used. Think of them as your emergency flare gun – they grab attention when you need it most.

Protocols for contacting authorities

- **Know the Numbers**: Make sure you know which numbers to call for different emergencies. It's like having the right keys for different locks.

- **Clear Communication**: When you contact authorities, be clear and concise. Who you are, where you are, what the situation is. It's like giving directions – the clearer, the better.

- **Backup Plans**: If phone lines are down, know the alternative ways to contact help. Could be ham radio, a neighbor, or even going there in person if safe to do so. It's like having a spare set of keys.

Networking with community safety groups

- **Join the Network**: Get involved with local community safety groups or neighborhood watches. It's like joining a club where everyone's got each other's backs.

- **Stay Engaged**: Attend meetings, participate in drills. It's not just about showing up; it's about being an active member. Think of it like being part of a sports team – teamwork makes the dream work.

- **Share Knowledge**: Exchange tips and information. Maybe you know the best way to secure windows, and your neighbor is a whiz at first aid. It's like a potluck dinner – everyone brings something to the table.

Reaching out for help in an emergency is all about having the right tools and knowing how to use them. Whether it's activating a beacon, making a distress call, or signaling to a neighbor, being prepared can make all the difference. And remember, in a community, safety is a team sport. The more you connect with others, the stronger your safety net becomes.

STOCKPILING SUPPLIES

"In the silence of a safe room, the well-prepared find a symphony of resources at their fingertips."

Essential Supplies Checklist

Water and food: storage and rotation

- **Water Wisdom**: Store enough water for at least several days. Think a gallon per person per day. It's like packing for a desert trek – you can never have too much water.

- **Food Fundamentals**: Stock non-perishable food items that are easy to prepare and consume. Canned goods, dried fruits, nuts – it's like your camping trip grocery list, but longer-lasting.

- **Rotation Routine**: Regularly rotate your supplies to keep them fresh. It's like managing a pantry – you don't want to find out your canned beans expired when it's too late.

First aid and medical necessities

- **First Aid Kit**: A well-stocked first aid kit is a must. Think bandages, antiseptics, pain relievers – it's your mini emergency room.

- **Prescriptions Prepared**: If anyone in your household needs prescription medication, keep an extra supply in your safe room. It's like having a spare tire – essential for those who need it.

- **Know Your Stuff**: Basic knowledge of first aid can be a lifesaver. Consider taking a course or at least familiarizing yourself with the contents of your kit. It's like knowing the rules of the road – important if you need to take the wheel.

Tools and miscellaneous must-haves

- **Tool Time**: Keep a set of basic tools – screwdrivers, a hammer, pliers. You never know when you might need to fix something on the fly.

- **Power Up**: Flashlights and extra batteries, or even better, hand-crank or solar-powered lights. It's like having candles and matches, but more reliable and longer-lasting.

- **Extra Extras**: Think about other essentials – blankets, a portable radio, spare clothes. Each item is like a piece in a puzzle – individually small, but together they create the bigger picture of preparedness.

Having a well-prepared safe room is like packing for an uncertain journey. You need enough supplies to sustain you, tools to aid you, and the knowledge to use them effectively. It's about thinking ahead, planning for the worst, and hoping you never have to use it.

***

Managing Your Inventory

Space-saving storage solutions

- **Think Vertical**: Use shelving to maximize vertical space. It's like building a skyscraper – going up instead of out. This way, you can store more without cluttering the floor space.

- **Containerize**: Use clear, stackable containers for organization. It's like playing Tetris with your supplies – everything fits neatly and is easy to find.

- **Hidden Spaces**: Utilize hidden or underutilized areas for storage – under beds, inside benches, or wall cavities. It's like finding secret compartments in a treasure hunt – every space has potential.

Keeping track of expiration dates and supply levels

- **Date Detective**: Regularly check and note the expiration dates on your supplies. It's like doing inventory in a store – you need to know what's still good and what needs replacing.

- **First In, First Out**: Rotate your stock based on expiration dates – use the oldest items first. It's like managing a kitchen pantry; you don't want things to go off.

- **Supply Spreadsheet**: Consider keeping a spreadsheet or a checklist to track what you have and what you need. It's like having a shopping list – it keeps you on track and organized.

Regularly updating your stockpile based on evolving needs

- **Review and Renew**: Regularly assess your stockpile and update it based on changing needs of your household. It's like updating your wardrobe – what worked last year might not be suitable this year.

- **Growth and Change**: If your family grows or someone's medical needs change, update your supplies accordingly. It's like adjusting your budget for a new expense or life event.

- **Seasonal Swaps**: Consider seasonal needs – like warmer clothing for winter or extra water in summer. It's like swapping out snow tires or packing for a holiday – you prepare for the conditions.

Managing your safe room inventory is all about organization, foresight, and adaptability. It's not just about having things; it's about having the right things at the right time and ensuring they're in good condition when you need them. Regular checks and updates are the key to keeping your safe room ready for anything.

***

Comfort and Morale

mportance of comfort items and entertainment

- **Personal Touches**: Just because it's a safe room doesn't mean it has to feel like a bunker. Bring in some comfort items like pillows, blankets, or even family photos. It's like having a little piece of home in a space meant for safety.

- **Entertainment Essentials**: Books, cards, board games, or even a tablet loaded with movies and games can be a lifesaver when you need to pass the time. It's like packing for a long flight – you want things to keep your mind occupied.

- **For the Little Ones**: If you have kids, include some of their favorite toys or coloring books. It's like bringing their favorite stuffed animal on a trip – a little piece of comfort goes a long way.

Psychological impact of color and design

- **Color Therapy**: Soft, calming colors can help reduce stress. Think pastels or earth tones. It's like designing a spa – you want a space that helps you unwind.

- **Light and Airy**: Make the space feel open and less claustrophobic. Use mirrors or light-colored decor to make the room feel larger. It's like choosing the right outfit to feel confident – the environment can boost your morale.

- **Organized and Clean**: A cluttered space can add to stress. Keep things organized and tidy. It's like having a clean desk – it sets the tone for a clear mind.

Personal items to include for emotional well-being

- **Sentimental Value**: Include items that have personal significance – a family heirloom, a favorite book, or a treasured gift. It's like carrying a piece of your history and memories with you.

- **Stress Relievers**: Think about what relaxes you. Maybe it's a scented candle (battery-operated for safety), a yoga mat, or a playlist of your favorite music. It's about creating a personal oasis.

- **Connection Items**: Items that remind you of the outside world and your loved ones, like letters, mementos, or even a digital photo frame loaded with pictures. It's like having a bridge to the world outside, keeping you connected to your life beyond the safe room.

Creating a space that's not only safe but also comforting is key to maintaining good morale during stressful situations. It's about making the best of a tough situation and providing a semblance of normality and calm in the midst of chaos.

Training and Drills

"A safe room's strength is tested not just by its walls, but by the readiness of those it shelters."

Understanding Your Safe Room

Familiarizing yourself with all features and functions

- **Know Your Fortress**: Spend time in your safe room to understand how everything works – from the ventilation system to the locks. It's like getting to know a new car – you need to know what every button does.

- **Practice Makes Perfect**: Regularly test and use the features. Whether it's the communication equipment or the emergency lighting, practice using them. It's like a fire drill – you want it to be second nature when you really need it.

- **Tech Savvy**: If you have advanced technology or gadgets in your safe room, make sure you know how to operate them. It's like having a smart home device – it's only useful if you know how to use it.

Creating a manual and guide for all inhabitants

- **Safe Room Bible**: Create a comprehensive manual that details every aspect of the safe room. Think of it as your safe room's instruction booklet.

- **Accessible Language**: Make sure the guide is understandable for all ages and literacy levels in your household. It's like writing instructions for a board game – clear and simple.

- **Emergency Procedures**: Include a section on what to do in different emergency scenarios. It's like having a playbook – everyone should know the plays.

Accessibility considerations for all ages and abilities

- **Universal Design**: Your safe room should be accessible to everyone in your household, regardless of age or ability. Think wider doorways for wheelchair access or non-slip flooring for safety.

- **Kid-Friendly Features**: If you have children, consider their needs – lower hooks for backpacks, reachable shelves, or even a designated space for them to feel comfortable. It's like setting up a kid's room – it needs to be functional for them.

- **Senior Support**: For elderly members, ensure ease of use – comfortable seating, easy-to-open doors, and clear, large-print instructions. It's like designing a space with your grandparents in mind – respectful, practical, and accessible.

Understanding your safe room is about more than just knowing where things are. It's about being intimately familiar with how everything works, ensuring everyone can use it effectively, and accommodating the needs of all household members. A well-understood safe room is an effective safe room.

***

Running Drills and Scenarios

Planning and executing effective drills

- **Schedule Regular Drills**: Just like a school fire drill, plan regular safe room drills. Set dates and times, and make sure everyone in the household knows when they'll happen. It's like setting a recurring meeting in your calendar – routine is key.

- **Realistic Scenarios**: Create scenarios that mimic potential emergencies – natural disasters, home invasions, etc. It's like role-playing – the more realistic, the better prepared you'll be.

- **Roles and Responsibilities**: Assign roles to each family member during the drill. This helps ensure everyone knows their part. It's like a sports team – everyone needs to know their position and play it well.

Simulating different threat levels and responses

- **Variety of Situations**: Plan for different levels of threats – from a power outage to a severe weather event. Each scenario might require a different response. It's like having a different game plan for different opponents.

- **Time Constraints**: Add elements like time limits to reach the safe room. This adds urgency and helps practice quick thinking and action. It's like a timed quiz – it tests your ability to act under pressure.

- **Unexpected Twists**: Introduce unexpected challenges during the drills, like a blocked path to the safe room. It's important to be adaptable. Think of it as an obstacle course – you need to be ready for anything.

Incorporating feedback to improve procedures

- **Debrief Post-Drill**: After each drill, gather everyone for a debrief. Discuss what went well and what could be improved. It's like reviewing game footage – you learn from what happened.

- **Continuous Improvement**: Take the feedback seriously

and make adjustments to your plan. It's an ongoing process, much like tweaking a recipe until it's just right.

- **Encourage Open Dialogue**: Make sure everyone feels comfortable sharing their thoughts and concerns. It's like having a family meeting – everyone's voice is important.

Running drills and scenarios for your safe room isn't just about going through the motions. It's about creating a realistic, adaptable, and continuously improving plan that will help keep you and your family safe in various emergency situations. Regular practice and open feedback are crucial for ensuring that everyone knows what to do when it really counts.

***

Mental Preparedness

Techniques for staying calm in high-stress situations

- **Breathing Exercises**: Just like in yoga or meditation, controlled breathing can be a powerful tool to calm the mind. Practice deep, slow breaths to help reduce anxiety and stress.

- **Mindfulness and Focus**: Engage in mindfulness practices. Focusing on the present moment can help prevent panic and maintain clarity. It's like using a magnifying glass – concentrate on one thing to see it clearly.

- **Positive Visualization**: Practice visualizing successful outcomes in stressful scenarios. It's like rehearsing a play in your mind where everything goes perfectly.

The role of leadership and organization in crisis

- **Clear Leadership**: Assign or take up the role of a leader in your safe room. Good leadership provides direction and can keep everyone focused and calm. Think of it as being the captain of a ship – guiding it through stormy seas.

- **Organized Approach**: Keep things organized. An organized space leads to an organized mind. It's like keeping your work desk tidy – it helps you think and act more clearly.

- **Decisive Actions**: In a crisis, indecision can be as dangerous as the wrong decision. Train yourself to make quick, informed decisions. It's like playing chess – think fast, but think ahead.

Coping strategies for extended stays

- **Routine is Key**: Establish a daily routine to bring a sense of normalcy. Regular meal times, sleep schedules, and even designated times for activities can help. It's like having a daily schedule – it adds structure to the day.

- **Stay Active**: Include physical activity in your routine. Exercise can be a great stress reliever and mood booster. It's like having a mini-gym session – a little can go a long way.

- **Emotional Outlets**: Have outlets for emotional expression – writing, drawing, or talking. Keeping a journal or having family discussions can be therapeutic. It's like having a safety valve – it helps release built-up pressure.

Mental preparedness is as crucial as physical preparedness in a safe room scenario. It's about training your mind to stay calm and clear, leading effectively, and maintaining psychological resilience, especially during extended stays. Techniques like controlled breathing, mindfulness, and maintaining routines can significantly impact your ability to cope effectively in high-stress situations.

BEYOND THE SAFE ROOM

"The true measure of safety is not just surviving the storm, but thriving after it has passed."

Reintegration and Recovery

Planning for after the crisis

- **Post-Crisis Plan**: Just like having a fire escape plan, have a plan for what to do once the crisis is over. Think about who to contact, where to go, and what steps to follow.

- **Stay Informed**: Keep updated on the situation outside. It's like checking the weather before heading out – you want to know what conditions you'll be stepping into.

- **Emotional Preparedness**: Understand that coming out of a crisis can be emotionally challenging. Prepare yourself and your family for a range of emotions. It's like coming back from a long trip – re-entry can be a shock to the system.

Steps for re-entering your home and neighborhood

- **Safety First**: Before re-entering your home or neighborhood, ensure it's safe to do so. Check for structural damage, gas leaks, or other hazards. It's like doing a safety check before driving a car – you need to ensure everything's in working order.

- **Neighborhood Network**: Connect with neighbors to get a sense of the broader situation. Share information and resources. It's like having a neighborhood meeting – everyone benefits from sharing knowledge.

- **Gradual Return**: Don't rush the process. Take gradual steps to reintegrate, especially if the crisis was prolonged. It's like dipping your toes in the water before diving in – take it slow.

Assessing and repairing damage

- **Damage Assessment**: Carefully assess any damage to your property. Document everything for insurance purposes. It's like being a detective at a crime scene – note every detail.

- **Seek Professional Help**: For significant repairs, consult professionals. Don't take on more than you can safely handle. It's like going to a doctor rather than self-diagnosing – expertise matters.

- **Mental Health Check**: Don't forget to assess and attend to your mental health. Seeking support from a counselor or therapist can be crucial. It's like healing a wound – sometimes you need professional help for proper healing.

Reintegration and recovery after using your safe room are about more than just physical rebuilding. It's a comprehensive process that involves safety checks, community coordination, and addressing both the physical and emotional aftermath. It requires careful planning, patience, and the willingness to seek help when needed. Remember, recovery is not just about fixing what's broken; it's about understanding and adapting to the new normal.

***

Community Building and Assistance

Leveraging your safe room for community support

- **Be a Resource**: Your safe room can be a valuable asset to your community. Offer to share it in times of need, like a neighborhood haven. It's like having a community center, but for safety.

- **Training and Workshops**: Host workshops or training sessions on safety and preparedness, using your safe room as a practical example. It's like giving cooking lessons in your kitchen – it's easier to learn where the action happens.

- **Fostering Community Ties**: Use your safe room project to foster stronger ties with neighbors. It's an opportunity to collaborate and strengthen community bonds. Think of it as a neighborhood project that brings everyone together.

Sharing resources and knowledge

- **Knowledge Exchange**: Share what you've learned about safety and preparedness with your community. Knowledge is a resource that becomes more valuable when shared. It's like teaching someone to fish – it empowers them.

- **Resource Pooling**: Collaborate with neighbors to pool resources for broader community safety. This could be shared storage of supplies or a network of safe houses. It's like creating a communal garden – everyone contributes and everyone benefits.

- **Open Communication Channels**: Establish open lines of communication for emergency situations. Having a neighborhood network can be invaluable in times of crisis. It's like having a neighborhood watch – everyone keeps an eye out for each other.

Establishing a neighborhood safety plan

- **Collective Planning**: Work with your community to create a neighborhood safety plan. Include meeting points, contact lists, and resource inventories. It's like drawing up a game plan for the whole team.

- **Drills and Simulations**: Organize neighborhood-wide drills. Practice makes perfect, and this holds true for community safety as well. It's like a fire drill at school – everyone needs to know what to do.

- **Local Partnerships**: Partner with local emergency services for training and advice. Their expertise can greatly enhance the effectiveness of your neighborhood plan. It's like having a guest coach for your local sports team.

Community building and assistance through your safe room is about turning individual preparedness into a collective strength. By sharing resources, knowledge, and a sense of responsibility, you not only enhance your own safety but also contribute to the resilience of your entire community. It's a powerful way of ensuring that when trouble comes, it's not just one house that's prepared, but the whole neighborhood.

***

Upgrades and Improvements

Incorporating new technologies and innovations

- **Tech Advancements**: Stay updated on new safety technologies and consider incorporating them into your safe room. It's like updating software on your computer – you want the latest and most efficient version.

- **Smart Systems**: Think about adding smart home technology for monitoring and controlling your safe room's environment. It's like having a high-tech butler – efficient and responsive.

- **Research and Development**: Regularly research new innovations in safety and security. Attend expos, read journals, or join online forums. It's like being a student of safety – always learning.

Learning from experience and adapting your safe room

- **Post-Use Review**: After using your safe room, take time to review what worked well and what didn't. It's like a sports team reviewing game tape – you want to see where you can improve.

- **Feedback Loop**: Encourage feedback from everyone who uses the safe room. Different perspectives can offer valuable insights. It's like having a brainstorming session – more ideas lead to better solutions.

- **Continual Evolution**: Make adjustments based on your experiences and feedback. A safe room should be a living project, adapting to changing needs. It's like evolving a recipe – tweaking it to make it just right.

Preparing for future threats with advanced planning

- **Anticipate Trends**: Keep an eye on global and local trends that might affect safety – climate change, urbanization, etc. It's like reading the weather forecast – you want to be prepared for what's coming.

- **Advanced Drills**: Regularly update and evolve your safety drills to reflect new threats or scenarios. It's like updating your workout routine – keeping it relevant to your current fitness goals.

- **Networking for Insight**: Engage with experts and authorities on safety and preparedness. Networking can provide early insights into emerging threats and responses. It's like having inside information – it gives you an edge.

Upgrades and improvements to your safe room should be an ongoing process. It involves staying informed about new technologies, learning from your experiences, and continuously adapting to new threats. Think of it as a journey rather than a destination – your safe room evolves as the world around it changes, ensuring you are always as prepared as possible for whatever may come.

CONCLUSION

Recap of Key Learnings

Summarizing the critical steps for building and maintaining a safe room

- **Solid Foundations**: Start with a thorough assessment – choose the right location, materials, and design for durability and practicality.

- **Security Essentials**: Focus on robust security features – from high-security doors and locks to sophisticated surveillance and communication systems.

- **Life Support Systems**: Ensure reliable life support – adequate ventilation, sustainable power supply, and efficient lighting solutions are crucial.

- **Supplies and Storage**: Stock up on essential supplies and manage your inventory with meticulous care – from food and water to medical supplies and tools.

- **Mental and Physical Preparedness**: Regular drills, mental preparedness strategies, and a clear understanding of your safe room's functionalities are vital for effective use.

- **Community Engagement**: Extend the concept of safety beyond your walls by engaging with and contributing to your community's safety and preparedness.

Reflecting on the importance of preparation and resilience

- **Preparation is Key**: Building and maintaining a safe room isn't just a physical task; it's a mindset. It's about anticipating needs and potential scenarios to ensure safety and security.

- **Resilience Through Adaptation**: The journey doesn't end once the safe room is built. Continuous learning, adapting to new threats, and technological advancements are essential for long-term resilience.

- **The Ripple Effect of Safety**: Your safe room represents a commitment not only to personal safety but also to the well-being of those around you. It's a testament to the importance of preparedness in fostering a resilient community.

In essence, the creation and maintenance of a safe room are about much more than construction and stocking up. It's a holistic approach to safety, encompassing physical security, mental preparedness, and community involvement. This journey emphasizes the importance of being proactive, adaptable, and ever-vigilant in the face of evolving threats and challenges.

***

The Broader Impact of Your Safe Room

The role your safe room plays in the safety of your family and community

- **Family Fortress**: For your family, the safe room is more than a physical space; it's a symbol of security and peace of mind. In times of uncertainty, it's a sanctuary that reassures and protects.

- **Community Beacon**: Beyond your family, your safe room can serve as a model and resource for the community. It showcases the importance of preparedness and can even provide refuge during communal emergencies.

- **Educational Tool**: Use your safe room as a practical example to educate others. It can spark conversations about safety and encourage others to take proactive steps in their own homes.

Encouraging continued education and preparedness

- **Lifelong Learning**: Emphasize the importance of staying informed about safety measures, new threats, and innovations in security and emergency preparedness. It's like keeping up with ongoing professional development – the learning never stops.

- **Sharing Knowledge**: Share your journey and experiences in building and maintaining the safe room with others. Workshops, community meetings, and social media are great platforms for this exchange.

- **Inspiring Action**: Your safe room can inspire others to think about their preparedness strategies. Lead by example and encourage neighbors, friends, and community members to consider their own safety measures.

The presence of a safe room in your home extends its impact far beyond its walls. It plays a crucial role in the safety and preparedness of not just your family but also your wider community. By serving as a beacon of safety and a hub of knowledge, it can inspire a culture of preparedness and resilience. This ongoing commitment to education and improvement highlights the dynamic nature of safety – it's always evolving and so should our strategies to ensure it.

***

FINAL THOUGHTS

As we reach the conclusion of our exploration into building and maintaining a safe room, it's crucial to remember that safety is a journey, not a destination. The creation of a safe room is just one step in an ongoing process of learning, adapting, and preparing for the unexpected. It's about cultivating a mindset that values vigilance, resilience, and community.

Safety as a Journey

- **Evolving Landscape**: The world we live in is ever-changing, and so are the challenges we face. Safety measures that are effective today may need to be adapted tomorrow. Treat safety as an evolving process, much like nurturing a garden – it requires continuous care and adaptation.

- **Proactive Mindset**: Being proactive about safety means always looking ahead, anticipating potential risks, and preparing accordingly. It's like playing chess – thinking several moves ahead.

- **Lifelong Commitment**: Commit to making safety a lifelong endeavor. It's not just about the physical structure of a safe room but also about building a culture of preparedness and awareness within your family and community.

Inviting Shared Experiences

- **Share Your Story**: I encourage you to share your experiences and insights. Whether it's the challenges you faced while building your safe room, the lessons you learned, or the sense of security it has brought to your life, your stories can be invaluable to others.

- **Learning from Each Other**: By sharing, we create a community of knowledge and support. Each story adds to a collective pool of wisdom that can help others in their journey towards safety and preparedness.

- **Building a Safety Network**: Your contributions can help foster a network of individuals committed to safety and resilience. This network can be a powerful resource for support, advice, and inspiration.

In closing, remember that the journey towards safety and preparedness is one we travel together. It's about more than just individual readiness; it's about building a community that stands strong in the face of adversity. Your safe room is a testament to that commitment, and your experiences are a valuable part of this collective journey. Share your story, learn from others, and continue to build on the foundation of safety you've established.

www.ingramcontent.com/pod-product-compliance
Lightning Source LLC
Chambersburg PA
CBHW061628130726
47996CB00003B/1171